101 Strawberry Recipes

by Carole Eberly

Cover by Kathi Terry

Illustrations by Gerry Wykes

Copyright 1986, 1994 by **eberly press**

1004 W. Michigan Avenue

E. Lansing, MI 48823

ISBN 0-932296-13-0

TABLE OF CONTENTS

Preserves, Jams, Sauces and Beverages 3

Breads and Salads 29

Dessert and Gooey Things 43

Preserves

Jams

Sauces and Beverages

SUNCOOKED STRAWBERRY PRESERVES

1/2 c. water
2 lbs. sugar
2 T. lemon juice
2 lbs. hulled strawberries

Bring water and sugar to a boil, stirring to dissolve sugar. Add lemon juice and strawberries. Simmer 2 minutes. Pour syrup and berries into large shallow pans. Set in the sun 3-4 days, until berries are plump and deep red, and syrup is jelled. Ladle into clean jars and seal. Makes 1 quart.

STRAWBERRY X BRUCE *HYBRID*

FOUR FRUIT CONSERVE

1 qt. hulled strawberries
1 qt. pitted tart cherries
1 sliced orange
2 c. chopped pineapple
Sugar
1 T. lemon juice
1/2 t. cinnamon

Mix all fruits together and weigh on scale. Add an equal amount of sugar and bring to a boil. Cook slowly until mixture is thick and clear. Ladle into jars and seal.

STRAWBERRY-HONEY JAM

    4 1/2 c. crushed strawberries
    7 c. mild honey
    1 3 1/2-oz. pkg. powdered fruit pectin

Crush strawberries with pectin. Stir constantly over high heat until mixture reaches full boil. Pour in honey and bring to a rolling boil. Boil and stir for 2 minutes. Remove from heat. Stir fruit and skim foam off for 5 minutes. Spoon into jars. Seal. Makes 11 6-oz. jars.

STRAWBERRY-RHUBARB JELLY

1 lb. diced rhubarb
Water
Juice & rind of lemon
Juice & rind of orange
1 qt. hulled strawberries
Sugar

Add lemon, orange juice and rinds to rhubarb in pot. Add enough water to prevent burning. Simmer until rhubarb is tender. Add strawberries and simmer 10 minutes. Strain juice through a jelly bag. Measure juice and bring to a boil. Stir in 1 cup sugar for each cup juice. Boil until mixture reaches jellying point, stirring often. Pour into jars and seal.

OLD FASHIONED STRAWBERRY JAM

8 c. crushed strawberries
6 c. sugar
3 T. lemon juice

Cook all ingredients slowly, stirring constantly until sugar dissolves. Boil until mixture reaches jellying point. Ladle into 6 half-pint jars and seal.

STRAWBERRY-RHUBARB JAM

3 c. mashed strawberries
3 c. diced rhubarb
6 c. sugar

Mix all ingredients together and boil for 8 minutes, stirring often to prevent sticking. Ladle into jars and seal. Makes 2 1/2 pints.

STRAWBERRY FREEZER JAM

    2 c. mashed strawberries
    4 c. sugar
    1 pkg. powdered fruit pectin
    1 c. water

Mix fruit and sugar in bowl; set aside for a half-hour. Boil pectin and water for 1 minute, stirring constantly. Remove from heat and add berries; stirring for 2 minutes. Pour into containers and cover. Chill until set. Store in freezer.

STRAWBERRY-CHERRY JAM

1-lb., 4-oz. can pitted tart cherries
10-oz. pkg. frozen strawberries
4 c. sugar
3 T. lemon juice
3 oz. liquid fruit pectin

Drain cherries. Measure enough cherries and juice to equal 2 c. Boil cherries, strawberries, sugar and lemon juice for 1 minute, stirring constantly. Remove from heat and add pectin. Alternately skim off foam and stir mixture for 5 minutes. Ladle into jars and seal. Makes 8 1/2 pt. jars.

FRESH STRAWBERRY SYRUP

    2 c. strawberries
    1/2 c. water
    1/2 c. sugar

Boil berries and water 10 minutes. Strain and sieve, saving juice and pulp. Add sugar to juice and pulp. Boil until thickened, about 8 minutes. Makes 1 cup.

GINGER-SPICED STRAWBERRIES

- 2 c. hulled strawberries
- 1 T. cornstarch
- 2 t. lemon juice
- 2 T. sugar
- 2 T. thinly sliced candied ginger
- 1/2 c. water

Mix all ingredients together in saucepan. Cook until clear and thick. If necessary, add more water. Serve cold with meats.

ORANGE STRAWBERRY TOPPING

4 c. hulled strawberries
1/2 c. sugar
1 T. grated orange rind
3/4 c. orange liqueur

Mix all ingredients in a 9x13" baking dish. Set aside for 30 minutes. Cover dish with foil and bake at 325 degrees for 40 minutes. Serve over ice cream or pound cake.

QUICK STRAWBERRY SYRUP

1 T. cornstarch
1/3 c. water
1/4 c. sugar
2 c. sliced strawberries
1 t. lemon juice

In a small saucepan, mix cornstarch and water. Add sugar and bring to a boil, stirring constantly, until mixture is thick and clear. Add strawberries and lemon juice. (Good on pancakes and waffles.)

STRAWBERRY HARD SAUCE

2/3 c. butter
2 c. powdered sugar
1 c. sliced strawberries

Cream butter and sugar. Stir in strawberries.

STRAWBERRY CREAM SAUCE
1 c. hulled strawberries
1/2 c. heavy cream

Mix strawberries and heavy cream in blender until smooth.

STRAWBERRY WHIP
1 c. sliced strawberries
1 c. sugar
1 egg white

Beat all ingredients in bowl for 10 minutes. Serve as cake topping.

STRAWBERRY CREAM FILLING
1/2 c. whipping cream
1/2 c. strawberry jam

Whip cream. Fold in jam. Spread between cake layers.

STRAWBERRY ICING I

2 c. mashed strawberries
1/4 c. powdered sugar
4 egg whites
Dash salt
1 t. vanilla
1/4 c. powdered sugar

Mix strawberries with powdered sugar and set aside. Beat egg whites with salt until stiff. Beat in vanilla and remaining powdered sugar. Drain strawberries. Fold pulp into mixture. Good on angel food cake.

STRAWBERRY ICING II

2 c. powdered sugar
Dash salt
1/4 c. hulled strawberries
1 t. strawberry liqueur

Mix everything together. If too thick, add a little more liqueur.

STRAWBERRY GLAZE

1 1/2 c. hulled strawberries
1 1/2 T. cornstarch
1/2 c. sugar

Mash strawberries with other ingredients in saucepan. Cook until thick and clear, stirring often. Cool slightly and pour over cake.

SEVEN-MINUTE STRAWBERRY FROSTING

2 egg whites
1 c. sugar

2/3 c. sliced strawberries
1/4 t. cream of tartar
Dash salt

In top of double boiler, beat all ingredients 1 minute. Place over boiling water, beating constantly until frosting forms peaks, about 7 minutes. Pour into mixing bowl and beat until of spreading consistency, about 2 minutes.

STRAWBERRY CREAM PUFF FILLING

1/2 pt. mashed strawberries
1/2 c. powdered sugar
Pinch salt
1 c. whipping cream
1/2 t. vanilla
12 cream puffs

Mix berries with powdered sugar and salt. Whip cream. Drain berries well; add pulp to whipped cream. Fold in with vanilla. Fill cream puffs.

FROZEN STRAWBERRY DAIQUIRIS

1 small can limeade concentrate
1 can water
1 c. rum
1 10-oz. pkg. sweetened frozen strawberries

Mix together thoroughly in a blender. Freeze until slushy.

IF STRAWBERRIES COULD DRAW...

STRAWBERRY CHAMPAGNE PUNCH

3 10-oz. pkg. frozen strawberries
1 large can pineapple chunks
3/4 c. sugar
2 bottles Sauterne
1 bottle Burgundy
2 bottles Champagne
1 bottle club soda

Mix strawberries, pineapple chunks, sugar and 1/2 bottle of Sauterne. Chill 8-10 hours. Pour fruit into punch bowl and pour chilled wines and club soda over. Serves a lot.

STRAWBERRY SHAKE

2 c. strawberries
1 1/2 c. milk
1/4 c. sugar
1 T. vanilla

Put all ingredients in blender and blast off. Serves 4.

SPARKLING STRAWBERRY PUNCH

3 10-oz. pkg. frozen strawberries
1/4 c. sugar
2 bottles Rose wine
2 6-oz. cans limeade concentrate
1 qt. soda water

Mix strawberries, sugar and 1 bottle of wine in punch bowl. Let stand for 1 hour. Combine mixture in blender with limeade. Pour back into punch bowl and add other bottle of wine and soda water. Add ice cubes or ice ring. Makes about 3 1/2 qts.

STRAWBERRY SHRUB

10-oz. pkg. frozen strawberries
3 c. water, divided
1/4 c. lime juice
3 T. lemon juice
1 T. grated lemon rind
1/4 c. sugar
1 c. orange juice

Boil berries and 2 cups water for 4 minutes. Press pulp through strainer. Add rest of water, along with remaining ingredients. Mix well; chill. Makes 1 quart.

STRAWBERRY LEMONADE

- 10-oz. pkg. frozen strawberries
- 1/4 c. sugar
- 2 c. water
- 3 c. lemonade

Boil berries, sugar and water 4 minutes. Strain and press pulp through a sieve. Add lemonade. Chill. Makes about 1 quart.

STRAWBERRY BLENDER SHAKE

1 c. boiling water
1 3-oz. pkg. strawberry gelatin
4 c. milk
1 qt. vanilla ice cream
Whole strawberries

Mix water and gelatin in blender until dissolved. Pour out half the mixture and reserve. Blend in half the milk and ice cream. Pour into glasses. Repeat with remaining ingredients. Garnish with strawberries.

STRAWBERRY TEA PUNCH

2 c. mashed strawberries
1 1/2 c. lemon juice
1 1/2 c. orange juice
1 1/4 c. sugar
1 c. water
6 c. weak tea

Mix berries and juices. Boil sugar and water to make a syrup. Add to berries with tea. Pour into punch bowl. Garnish with strawberry halves. Makes 2 1/2 qts.

OLD FASHIONED STRAWBERRY ICE CREAM SODA

1/3 c. sliced strawberries
2 T. sugar
4 T. milk
Strawberry ice cream
Carbonated water

Mix strawberries, sugar and milk in a blender. Fill a tall glass half full of carbonated water. Stir in strawberry mixture and top with a scoop of ice cream. Grab a straw and the nearest hammock.

STRAWBERRY EGG NOG

    1 c. strawberries
    1 c. milk
    1/4 c. sugar
    2 eggs
    Dash cinnamon
    2 t. vanilla

Mix all ingredients together in blender. (A quick breakfast.)

Breads and Salads

STRAWBERRY JAM BREAD

1 egg
1/4 c. sugar
2 T. melted butter
1 c. strawberry jam
2 c. flour
1 T. baking powder
1/4 t. baking soda
1/2 t. salt
1/2 c. milk

Beat egg, sugar and melted butter. Mix in jam. Sift dry ingredients and blend into creamed mixture alternately with milk. Pour into well-greased loaf pan and bake at 350 degrees for 1 hour.

STRAWBERRY BISCUITS

2 c. flour
1 t. salt
1 T. baking powder
1/4 t. baking soda
1/4 c. shortening
1 c. buttermilk
1/2 c. melted butter
24 small sugar cubes
24 strawberries

Sift dry ingredients together. Mix in shortening. Blend in buttermilk. Knead on heavily floured board until soft and spongy. Roll to 1/2-inch thickness. Cut out biscuits. Place on well-greased sheet and brush tops with melted butter. Press a sugar cube and a strawberry, side by side, in center of each. Let biscuits raise 15 minutes. Bake at 450 degrees for 15 minutes.

STRAWBERRY BREAD

1/2 c. butter
1 c. sugar
1/2 t. almond extract
2 eggs, separated
2 c. flour
1 t. baking powder
1 t. baking soda
1 t. salt
1 c. crushed strawberries
1 T. sugar

Cream butter, sugar and extract. Beat in egg yolks. Sift flour, baking powder, soda and salt. Mix into creamed mixture alternately with strawberries. Beat egg whites until stiff. Fold into batter. Pour into a greased loaf pan. Sprinkle sugar on top. Bake at 325 degrees for 50-60 minutes.

STRAWBERRY-FILLED PANCAKES

1 1/4 c. flour
1 1/2 c. milk
5 eggs
1/4 c. melted butter
1/2 t. salt
3 T. sugar
3 c. sliced strawberries
1/2 c. sugar
Powdered sugar
Whipped or sour cream

Make a batter from the flour, milk, eggs, butter, salt and 3 T. sugar. In a separate bowl, mix together strawberries and 1/2 c. sugar. In a large skillet, melt 1 T. butter for each pancake. Add 1/6 of the batter and cook on both sides until stiff. Place 1/2 c. strawberries on edge of each pancake and roll up. Sprinkle with powdered sugar. Serve with whipped or sour cream. Makes 6.

STRAWBERRY FRITTERS

2 c. firm chopped strawberries
1 c. flour
1 T. sugar
1 t. baking powder
1/4 t. salt
1/2 c. milk
1 egg
Powdered sugar

Drain strawberries well. Sift dry ingredients into bowl. In other bowl, mix milk and egg; add to dry ingredients with strawberries. Drop batter by teaspoonful into hot cooking oil. Fry until golden brown. Drain on paper towels. Sprinkle with powdered sugar. Makes about 30.

STRAWBERRY PECAN BREAD

1 c. butter
1 1/2 c. sugar
1 t. vanilla
1/2 t. lemon extract
4 eggs
3 c. sifted flour
1 t. salt
1 t. cream of tartar
1/2 t. baking soda
1 c. strawberry jam
1/2 c. sour cream
1 c. chopped pecans

Cream butter, sugar, vanilla and extract. Beat in eggs. Sift together flour, salt, cream of tartar and soda. Combine jam and sour cream. Add jam mixture alternately with dry ingredients to creamed mixture. Stir in nuts. Bake in greased loaf pans at 350 degrees for 50-55 minutes.

DON'T LET THE SUN FOOL'YA – IT TURNS YOU RED & THEN THEY GET YOU!

STRAWBERRY SOUR CREAM MOLD

3-oz. pkg. strawberry gelatin
1 T. unflavored gelatin
2 c. boiling water
2 T. lemon juice
2 10-oz. pkg. frozen strawberries
1 c. sour cream

Mix gelatins together. Add boiling water, stirring until dissolved. Mix in lemon juice and strawberries. Chill until thickened. Pour half mixture into a 5-cup mold. Spoon sour cream gently on top of gelatin. Spoon remaining gelatin on top. Chill. Serves 8.

FROZEN STRAWBERRY-PECAN MOLD

1 6-oz. pkg. strawberry gelatin
1 1/2 c. boiling water
1 10-oz. pkg. frozen strawberries
1 small can crushed pineapple
1/2 c. chopped pecans
2 diced bananas
1/2 pt. sour cream
1 t. unflavored gelatin

Dissolve gelatin in water. Mix in remaining ingredients, except sour cream and gelatin. Pour half mixture in 9x9-inch pan. Chill until set. Mix sour cream and gelatin; let stand 5 minutes. Spread over top. Smooth remaining gelatin mixture on top. Chill until set.

STRAWBERRY FRUIT FROST

12 ozs. orange juice concentrate
3 c. water
1 c. sugar, dissolved in 1 c. water
1 large can crushed pineapple, undrained
1 10-oz. pkg. frozen strawberries
4 sliced bananas

Mix all ingredients in a 9x13-inch pan. Freeze. Remove 15 minutes before serving.

FRESH FRUIT SALAD

2 qts. strawberries
2 c. green grapes
1 pineapple, chunked
2 c. Mandarin oranges, with juice
2 bananas
Lemon juice

Mix all ingredients except bananas and lemon juice. Just before serving, slice bananas and dip in lemon juice. Mix with other fruit. Serve with FRUIT SALAD DRESSING, if desired.

FRUIT SALAD DRESSING

2 eggs
3/4 c. sugar
3 T. lemon juice
1/2 pint whipping cream

Beat eggs, sugar and lemon juice together. Cook over low heat until thickened. Cool. Whip cream and fold into dressing.

CRAN-BERRY SALAD

1 3-oz. pkg. strawberry gelatin
1 3-oz. pkg. lemon gelatin
1 1/2 c. boiling water
10-oz. frozen strawberries
16-oz. can cran-raspberry sauce
7-oz. bottle 7-up

Dissolve gelatins in boiling water. Stir in strawberries. Stir cran-raspberry sauce into gelatin mixture. Chill until partially set. Slowly pour in 7-up, stirring gently. Turn into 6-cup mold and chill.

CHILLED STRAWBERRY SOUP

1 qt. strawberries
1/2 c. water
1/2 c. orange juice
1/2 c. sugar
2 c. yogurt

Mix all ingredients in blender. Chill. Serves 6.

SUNRISE MOLD

1 T. unflavored gelatin
1/3 c. cold water
1 3-oz. pkg. strawberry gelatin
1 c. boiling water
10-oz. pkg. frozen strawberries
1 small can crushed pineapple
1 3-oz. pkg. softened cream cheese
1/2 c. chopped pecans
1/2 pint whipping cream, whipped

Soften unflavored gelatin in cold water. Dissolve both gelatins in boiling water. Add strawberries, pineapple and cream cheese. Fold in nuts and whipped cream. Pour into mold. Chill until firm. Serves 8.

Desserts

and

Gooey Things

STRAWBERRY CAKE

1 c. flour
1/2 c. sugar
2 t. baking powder
Dash salt
1 egg
2 T. melted butter
1/2 c. milk
1 1/2 c. firm strawberries

Sift together flour, sugar, baking powder and salt. Add egg, butter and milk. Pour into a greased 8-inch square pan. Top with sliced strawberries. Sprinkle with topping of 1/2 c. flour, 1/2 c. sugar, 1/4 c. butter and 1/4 c. chopped nuts. Bake at 375 degrees for 35 minutes.

STRAWBERRY DUMP CAKE
(An easy cake for the gang - and one that quickly will vanish.)

 1 15 1/2 oz. can crushed pineapple & juice
 1 c. coconut
 1 can strawberry pie filling
 1 box yellow cake mix
 3/4 c. butter
 1 c. chopped pecans

Spread pineapple on bottom of 9x13-inch pan. Pour pie filling on top. Sprinkle coconut over top, followed by cake mix. Slice butter patties on top. Sprinkle pecans over mixture. Bake at 350 degrees for 45 minutes.

STRAWBERRY-RHUBARB CAKE

1 1/2 c. brown sugar
1/2 c. butter
1 egg
1 c. sour cream
2 c. flour
1 t. baking soda
1 t. vanilla
1 c. diced rhubarb
1 c. sliced strawberries
2 T. brown sugar
1 t. cinnamon

Cream sugar and butter; add egg. Sift flour with soda; add alternately with sour cream to sugar mixture. Stir in vanilla, rhubarb and strawberries. Sprinkle with brown sugar and cinnamon. Pour in greased 9x13-inch pan. Bake at 350 degrees 40-45 minutes.

STRAWBERRY TORTE

1 c. strawberries	1/2 t. cream of tartar
1/4 c. strawberry jam	2 c. sugar
Lemon juice	1 t. vanilla
8 egg whites	1 t. vinegar
1/2 t. cream of tartar	whipped cream

Stir jam, thinned with lemon juice, into strawberries; set aside. Beat egg whites with cream of tartar until very stiff. Slowly add 1 c. sugar. Beat in vanilla and vinegar. Gradually add last cup sugar, beating until very stiff. Bake at 250 degrees in a greased spring form pan for about 1 hour, 15 minutes. Cool and garnish with whole drained strawberries.

STRAWBERRY-CHERRY COBBLER

21-oz. can strawberry pie filling
10-oz. pkg. frozen raspberries, thawed & drained
2 t. lemon juice
1/2 c. flour
1/4 c. sugar
1/4 c. butter
Dash salt

Mix pie filling, raspberries and lemon juice in saucepan; bring to a boil. Pour into greased 8-inch square baking pan. Mix remaining ingredients until crumbly. Sprinkle over filling. Bake at 375 degrees for 30 minutes.

STRAWBERRY SHORTCAKE

1 qt. sliced strawberries	3 t. sugar
1 c. sugar	1/3 c. shortening
1 c. flour	3/4 c. milk
2 t. baking powder	Soft butter
1 t. salt	Whipped cream

Mix strawberries and sugar; set aside. Sift dry ingredients together. Cut in shortening; blend until crumbly. Make a well and pour in milk. Form dough into a ball. Knead on lightly floured board about 15 times. Pat dough to 1/2-inch thickness. Cut into 3-inch rounds. Spread half the rounds with soft butter. Top with remaining rounds. Bake on baking sheet at 450 degrees for 10-15 minutes. Split shortcakes and spoon half the strawberries over bottom. Top with shortcakes and more strawberries. Top with whipped cream.

STRAWBERRY CAKE

1 pt. hulled strawberries
1/2 c. sugar
1 pkg. yellow cake mix
1 3-oz. pkg. strawberry gelatin
3/4 c. cooking oil
1 c. chopped nuts
4 eggs
2 T. flour

Mix together strawberries and sugar; set aside. Beat remaining ingredients. Fold in strawberry mixture. Bake in 9x13-inch greased pan at 350 degrees for about 45 minutes. Serve with whipped cream.

MICHIGAN STRAWBERRY JAM CAKE

- 1 c. buttermilk
- 1 t. baking soda
- 1 c. butter
- 2 c. sugar
- 3 c. flour
- 1 t. nutmeg
- 1/2 t. cinnamon
- 1/4 t. cloves
- 6 eggs
- 1 1/2 c. strawberry jam

Mix buttermilk and soda together; set aside. Cream butter and sugar. Sift together flour and spices. Add alternately to creamed mixture with buttermilk and eggs. Mix in jam. Bake in greased bundt pan at 350 degrees for about 45 minutes. Cool and cover with STRAWBERRY GLAZE, if desired.

STRAWBERRY BETTY

1 qt. hulled strawberries
1 T. lemon juice
2/3 c. brown suga
r4 c. white bread cubes
1/4 c. white sugar
2 T. butter

Mix berries with lemon juice and brown sugar. Put in 8-inch square baking pan. Mix bread cubes with sugar and sprinkle over strawberries. Dot with butter. Bake at 350 degrees for 25-30 minutes. Serve warm with cream.

STRAWBERRY JAM CAKE

1 c. brown sugar
1/2 c. butter
1 c. strawberry jam
1 c. smashed bananas
2 cups flour
1/4 t. cinnamon
1/4 t. nutmeg
Pinch of cloves
1 t. baking soda
1/2 c. sour cream
2 eggs

Cream sugar and butter. Stir in jam and bananas. Sift dry ingredients; add alternately with milk to creamed mixture. Beat in eggs. Pour into a greased 13x9-inch pan. Bake at 350 degrees 40 minutes. Cool. Dust with powdered sugar or frost with favorite icing.

EASY STRAWBERRY CREAM PIE

1 lg. can evaporated milk
10-oz. pkg. frozen strawberries
1 3-oz. pkg. strawberry gelatin
1/4 c. sugar
9-inch graham cracker pie shell

Chill evaporated milk in refrigerator with beaters and bowl. Bring remaining ingredients to a rolling boil. Cool. Whip milk until thick. Fold in strawberry mixture. Pour into pie shell. Chill.

STRAWBERRY CHIFFON PIE

4 eggs, separated
1 c. sugar, divided
1 10-oz. pkg. frozen strawberries
1 envelope unflavored gelatin
1/2 c. cold water
9" graham cracker pie shell
1 pt. whipping cream, whipped

Beat yolks with 1/2 c. sugar until light. Add strawberries. Cook in top of double boiler until mixture coats spoon. Remove from heat; add gelatin after soaking in 1/2 c. cold water. Cool. Beat egg whites until stiff, slowly adding remaining sugar. Fold into berry mixture. Pour into pie shell. Chill. Top with whipped cream.

STRAWBERRY-TOMATO PIE

10-inch unbaked pie shell with lattice top
1 qt. strawberries
1 c. sugar
4 peeled and cubed tomatoes
6 T. flour
1/4 t. cinnamon
3 T. sugar
3 T. butter

Mix strawberries and sugar; let stand 1 hour. Pour half the strawberries and juice into pie shell. Sprinkle with half the flour and cinnamon. Top with half the tomatoes and sugar. Dot with half the butter. Repeat layers. Cover with lattice top. Bake at 450 degrees for 15 minutes. Lower heat to 400 degrees for 25 more minutes. Chill.

STRAWBERRY-CHEESE PIE

Baked 9-inch pie shell
1 qt. sliced strawberries
3-oz. pkg. soft cream cheese
1 1/2 c. strawberry juice
1 c. sugar
3 T. cornstarch

Spread cream cheese on bottom of pie shell. Cover with half the berries. Mash and strain rest of berries, extracting the juice. Bring juice to a boil. Stir in sugar and cornstarch. Cook, stirring constantly, until boiling. Boil 1 minute. Pour over berries in pie shell. Chill 2 hours. Serve with whipped cream.

LAZY DAY STRAWBERRY CREAM PIE

1 pt. sliced strawberries
1/4 c. sugar
1 3-oz. pkg. strawberry gelatin
1 c. boiling water
1 pt. softened vanilla ice cream
Baked 9-inch pie shell

Mix strawberries and sugar; set aside. Dissolve gelatin in boiling water. Blend in ice cream. Cool until slightly thickened. Drain berries and add to gelatin mixture. Pour into pie shell. Chill.

FAVORITE STRAWBERRY PIE

Baked 9-inch pie shell
3 c. sliced strawberries
3/4 c. water
3 T. cornstarch
1 c. sugar
1 t. lemon juice

Cook 1 c. strawberries in water for 3-4 minutes. Combine cornstarch and sugar; add lemon juice, and cook with strawberries until mixture is thick and clear. Cool. Pour 2 c. berries in pie shell. Pour sauce over. Chill 1-2 hours. Top with whipped cream.

FORTH OF JULY PIE

Baked 9-inch graham cracker pie shell
2 eggs
1 15-oz. can sweetened condensed milk
1/2 t. lemon peel
1/2 c. lemon juice
2 T. powdered sugar
1/2 pt. whipped cream
1 pt. whole strawberries, hulled
1 c. blueberries

Beat together eggs, milk and lemon peel. Slowly add lemon juice. Continue beating until mixture is thickened. Pour into pie shell. Chill 2-3 hours. Beat whipping cream with powdered sugar. Spread on top of pie. Make a row of upstanding strawberries around rim of pie. Scatter blueberries in center. Refrigerate.

STRAWBERRY-MARSHMALLOW PIE

Baked 9-inch pie shell
4 c. miniature marshmallows, divided
3/4 c. milk
1 c. whipping cream, whipped
1 t. vanilla
1-lb., 5-oz. can strawberry pie filling

Melt 3 1/3 c. marshmallows in milk, stirring constantly. Chill until partially set. Fold in whipping cream and vanilla. Pour half the mixture into pie shell. Spoon in pie filling. Fold remaining marshmallows into cream mixture and spread over top. Garnish with fresh strawberries.

STRAWBERRY TARTS

6 baked tart shells
2 T. cornstarch
1 c. sugar
1 qt. sliced strawberries
1 T. butter
2 t. lemon juice
3-oz. pkg. strawberry gelatin

Mix cornstarch, sugar, strawberries and sugar in saucepan. Stir until boiling for 2 minutes. Remove from heat and add remaining ingredients, stirring until gelatin dissolves. Pour into shells. Chill.

STRAWBERRY BITES

1 1/2 c. powdered sugar
1 c. butter
1 egg
1 t. vanilla
1/2 t. almond extract
2 1/2 c. flour
1 t. baking soda
1 t. cream of tartar
1 1/2 c. strawberry jam
Ground walnuts

Beat together powdered sugar, butter, egg, vanilla and extract. Stir in flour, baking soda and cream of tartar. Cover and refrigerate until dough is firm. Separate dough into 12 equal parts, shaping each into a strip, 8x1 1/2-inches on greased cookie sheets. Take a wooden spoon and make a lengthwise indentation down the center of each strip. Spread 1 teaspoon jam in each. Sprinkle with ground nuts. Bake at 375 degrees for 10 minutes. Cut strips into 1" diagonal pieces. Makes about 8 dozen.

STRAWBERRY CHEESECAKE COOKIES

2 8-oz. pkgs. cream cheese
2 eggs
3/4 c. sugar
1 T. vanilla
16 vanilla wafers
Strawberry jam

Beat cream cheese, eggs, sugar and vanilla. Place one vanilla wafer in a muffin paper in muffin tins. Pour mixture to 3/4 full in each paper. Bake at 375 degrees 10 minutes. Cool. Top each with a dab of strawberry jam.

STRAWBERRY RIBBON BARS

1/4 c. melted butter
1 c. strawberry jam
4 eggs
3/4 c. sugar
2 t. vanilla
3/4 c. flour
1 t. baking powder
1 t. salt
1/2 t. cinnamon

Mix butter and jam together; spread on bottom of 5 1/2 x 10 1/2-inch pan. Beat eggs and sugar together; add vanilla. Sift dry ingredients and fold into egg mixture. Spread over jam and bake at 400 degrees for 15-18 minutes. Remove from oven and let stand 5 minutes. Invert pan on sheet of waxed paper, lightly dusted with powdered sugar. After 2-3 minutes, lift pan gradually, letting cake fall out. Cut cake crosswise. Put two jam sides together. Cut into 2" x 1 1/2" bars.

STRAWBERRY OATMEAL BARS

3/4 c. butter
1 c. brown sugar
1 3/4 c. flour
1/2 t. baking soda
1 t. salt
1 1/2 c. rolled oats
1 1/2 c. strawberry jam

Cream butter and sugar. Sift together flour, soda and salt. Add sugar mixture, stir in oats. Press half the mixture on bottom of greased 9x13-inch pan. Spread with jam. Cover with remaining mixture, pressing gently. Bake at 400 degrees for 25 minutes. Makes about 2 dozen.

STRAWBERRY-COCONUT BARS

1 c. flour
1/2 c. butter
3 T. powdered sugar
2 eggs
1 c. sugar
1/4 c. flour
1 t. vanilla
1/2 t. baking powder
1/8 t. salt
1/2 c. chopped walnuts
1/2 c. coconut
1/2 c. strawberry jam

Combine 1 c. flour, butter and powdered sugar. Press in greased 8-inch square baking pan. Bake at 350 degrees for 10 minutes. Beat eggs. Stir in remaining ingredients. Spread over baked layer. Bake at 350 degrees 25-30 minutes. Cool. Makes about 24.

STRAWBERRY THUMBPRINT COOKIES

1/4 c. brown sugar
1/2 c. butter
1 egg, separated
1/2 t. vanilla
1 c. flour
1/8 t. salt
3/4 c. ground walnuts
Strawberry jelly

Cream brown sugar and butter. Mix in egg yolk and vanilla. Stir in flour and salt. Roll dough into 1" balls. Dip balls into egg white; roll in nuts. Place on ungreased cookie sheet. Press thumb deep in center of each. Bake at 350 degrees until light brown, about 10 minutes. Cool. Fill centers with jelly. Makes about 36.

STRAWBERRY FILLED COOKIES

2 c. brown sugar
1 c. shortening
2 eggs
1/2 c. milk
1 t. vanilla
3 1/2 c. flour
1 t. salt
1 t. baking soda
Strawberry jam

Beat brown sugar, shortening and eggs. Stir in milk and vanilla. Sift together flour, salt and baking soda. Combine into mixture. Drop by teaspoonsful onto ungreased cookie sheet. Top each spoonful of dough with 1/2 teaspoon strawberry jam. Put another 1/2 teaspoon dough on top of jam. Bake at 400 degrees for 10-12 minutes. Makes about 60 cookies.

STRAWBERRIES & MERINGUES

2 c. sliced strawberries
2 T. sour cream
1 t. strawberry flavored liqueur
4 T. powdered sugar
2 egg whites
1/8 t. cream of tartar
8 T. sugar
1/4 t. almond extract
1 t. cocoa

Mix strawberries, sour cream, liqueur and powdered sugar together; set aside. Beat egg whites with cream of tartar until soft peaks form. Add sugar slowly, beating well after each addition. Mix in almond extract and cocoa, beating until stiff peaks are formed. Drop meringues in 6 portions on greased cookies sheets. Make a nest in each with back of a spoon. Bake at 275 degrees for 30 minutes. Turn off oven and leave in another 20 minutes. Cool. Fill centers with strawberry mixture. Makes 6.

FROZEN STRAWBERRY CREAM

1 c. flour
1/2 c. melted butter
1/2 c. brown sugar
1/2 c. chopped nuts
2 egg whites
2/3 c. sugar
10-oz. pkg. frozen strawberries
1 T. lemon juice
1 c. whipping cream
2 T. sugar

Mix flour, butter, brown sugar and nuts. Press into a 9x13-inch pan. Bake at 250 degrees for 20 minutes. Crumble 2/3 of mixture into 9x13-inch pan, press down. Whip egg whites, sugar, strawberries and lemon juice at high speed for 10 minutes. Fold in cream whipped with 2 T. sugar. Spoon over crumbs. Sprinkle remaining crumbs on top. Freeze at least 8 hours. Serves 15.

STRAWBERRIES SUPREME

1 qt. hulled strawberries
1/4 c. cherry liqueur
1 t. sugar
1 10-oz. pkg. raspberries
1/2 pt. whipping cream, shipped
1 T. powdered sugar

Mix strawberries, liqueur and sugar. Puree raspberries in blender; strain, Pour over strawberries and stir. Chill. Whip cream with powdered sugar and spoon over strawberries.

STRAWBERRIES IN THE SNOW

1 3-oz. pkg. cream cheese
1/2 c. sugar
2 t. lemon juice
1 t. vanilla
1/2 pt. whipping cream, whipped
1 can strawberry pie filling
9-inch graham cracker pie shell

Beat cream cheese, sugar, lemon juice and vanilla until fluffy. Fold in whipping cream. Spread in pie shell. Cover with pie filling. Chill 2 hours before serving.

STRAWBERRIES & ORANGES

6 oranges, peeled
2 qts. hulled strawberries
4 T. sugar

Section oranges; combine with strawberries and sugar.

STRAWBERRY RUSSIAN CREAM

1 pt. sliced strawberries
1/4 c. sugar
1 c. whipping cream
3/4 c. powdered sugar
1/2 c. cold water
1 1/2 t. gelatin
1 c. sour cream
1 t. vanilla

Mix strawberries and sugar; set aside. Add powdered sugar to whipping cream and heat in top of double boiler. Meanwhile, soak gelatin in cold water. When cream mixture is lukewarm, add gelatin and stir until gelatin is dissolved. Remove from heat and cool. When it begins to thicken, fold in sour cream. Beat mixture well, adding vanilla. Pour into 8-inch square pan and cut into squares. Serve covered with strawberry mixture.

FROSTY STRAWBERRY DESSERT

1 c. flour
1/4 c. brown sugar
1/2 c. chopped walnuts
1/2 c. melted butter
2 egg whites
3/4 c. sugar
2 c. sliced strawberries
2 T. lemon juice
1/2 t. vanilla
1 c. whipping cream, whipped

Mix flour, brown sugar, nuts and butter; spread in a 9x13-inch pan. Bake at 350 degrees for 20 minutes. Cool. Remove 1/3 c. of mixture for topping. Beat egg whites, sugar, berries and lemon juice until mixture begins to thicken. Add vanilla. Beat until stiff peaks form, about 10-12 minutes. Fold in whipped cream. Spoon into pan. Top with reserved crumbs. Freeze at least 6 hours. Serves 12-15.

STRAWBERRY & FRUIT CASSEROLE

1 lb. peaches, peeled & halved
1 lb. pears, peeled & halved
1 1/2 c. strawberry jam
Brown sugar
1/4 c. dark rum
Sour cream

Spread jam in bottom of 2 qt. baking dish. Arrange fruit on top and sprinkle with brown sugar to taste. Pour rum over top. Bake at 450 degrees for 15-20 minutes. Serve with sour cream.

TIPSY FRUIT MEDLEY

3 pears, peeled & cored
2 oranges, peeled & sliced
1 c. halved green grapes
2 c. sliced strawberries
2 bananas, peeled & sliced
1/2 c. orange juice
1/2 c. dry white wine

Cut pears into bite size pieces. Cut orange slices in half. Mix all fruit together. Combine orange juice and wine. Pour over fruit, mixing until fruit is well-coated.

STRAWBERRY-RHUBARB DELIGHT

3 lb. rhubarb, cut in 1-inch pieces
1/8 t. cinnamon
1/3 c. sugar
1 c. water
Water
1 T. cornstarch
1/4 c. cold water
1 t. lemon juice
2 c. sliced strawberries

Mix and bring to a boil the rhubarb, cinnamon, sugar and 1 c. water. Simmer until tender, about 3 minutes. Remove from heat and drain, reserving syrup. Add water to equal 1 1/4 c. syrup. Mix cornstarch with cold water until dissolved. Add to syrup mixture and cook, stirring until thick. Cook 2 minutes more. Remove from heat. Stir in lemon juice and strawberries. Chill. Serve with whipped cream. Makes 8 servings.

DOUBLE STRAWBERRY BAKED ALASKA

1 1/2 c. sliced strawberries
1/4 c. sugar
2 egg whites
1/4 t. cream of tartar
8 T. sugar
6 individual sponge cake shells
Strawberry ice cream

Mix strawberries with sugar and set aside. Beat egg whites with cream of tartar until foamy. Slowly add sugar, beating until egg whites hold a stiff peak. Drain strawberries before putting a tablespoon or so on each cake. Top with a scoop of ice cream. Cover whole cake with meringue, sealing edges. Bake at 450 degrees for 3 minutes, or until golden brown.

KIRSH STRAWBERRIES IN CUSTARD SAUCE

2 qts. hulled strawberries
2 c. strawberry jam
1/4 c. Kirsh
1 c. heavy cream
3 T. sugar
4 egg whites
1 c. heavy cream
2 t. vanilla

Mix strawberries, jam and Kirsh together. Chill. In top of double boiler, scald 1 c. heavy cream. Stir in sugar. Beat egg whites with other cup of heavy cream. Stir a little of the hot mixture into the egg whites. Gradually add the egg whites to the hot cream. Cook mixture, stirring constantly, over boiling water. When thick, remove from heat and stir in vanilla. Cool. Pour over berries.

HOT STRAWBERRIES IN WINE SAUCE

    4 c. water
    1 3/4 c. sugar
    1 t. lemon juice
    Dash vanilla
    3 pt. hulled strawberries
    4 egg yolks
    4 T. sugar
    4 oz. sweet red wine

Boil water, sugar, lemon juice and vanilla about 5 minutes. Add berries and boil 1 more minute. Drain berries in colander. Spread in shallow glass serving bowl. Beat egg yolks, sugar and wine until mixture is foamy and peaks. Spread over hot strawberries and serve. Serves 6.

HOW STRAWBERRIES REALLY SPREAD.

STRAWBERRY RIFFLE

1 T. unflavored gelatin
1/3 c. cool water
1 c. cottage cheese
10-oz. pkg. frozen strawberries

Stir gelatin in water in heat-proof measuring cup. Put in pan with 1-inch simmering water. Put the cottage cheese in a blender, adding some liquid from the berries. When gelatin is dissolved, put into the cottage cheese mixture along with the berries. Blend. Chill. Serves 4.

STRAWBERRY YOGURT DESSERT

2 1/2 c. hulled strawberries
1/3 c. whipped dessert topping
1/3 c. plain yogurt
1 T. vanilla
1 T. sugar

Crush 1/3 c. strawberries; halve remaining ones. Mix remaining ingredients together with crushed berries. Spoon halved berries into 4 serving dishes. Top with yogurt mixture.

STRAWBERRIES & SOUR CREAM

2 pts. hulled strawberries
1 pt. sour cream
1/2 c. brown sugar

Place strawberries, sour cream and brown sugar into separate bowls. Have guests dip strawberries into sour cream and then into sugar.

STRAWBERRY SORBET

2 qts. hulled strawberries
1 c. sugar
1 c. water
1 T. lemon juice

Puree berries in blender. Boil sugar and water for 5 minutes. Cool. Mix in blender with berries and lemon juice. Freeze until mushy in refrigerator trays, stirring occasionally. Serves 4-6.

STRAWBERRY FOOL

2 c. hulled strawberries
1/3 c. sugar
1/4 c. water
2 t. lemon juice
1 c. whipping cream, whipped

Bring to a boil the berries, sugar and water. Reduce heat; simmer for 10 minutes. Stir in lemon juice and cool. Puree berries in blender. Spoon whipped cream into serving bowl. Pour puree over cream. With a spatula, run through the mixture several times to marbelize. Cover and chill. Serves 6.

STRAWBERRY JULIET

3 pts. hulled strawberries
1/2 c. sugar
1 pt. vanilla ice cream
1/2 c. whipping cream
2 t. lemon juice
1/4 c. orange liqueur

Mix strawberries with sugar; set aside. Stir ice cream slightly, fold in whipped cream. Stir in lemon juice and orange liqueur. Pour over berries in serving dishes or large bowl.

WINTER STRAWBERRIES

1 6-oz. pkg. strawberry gelatin
1 c. ground pecans
1 c. ground coconut
3/4 c. sweetened condensed milk
1 t. vanilla
Red sugar crystals

Green food color
Slivered almonds

Mix together gelatin, pecans, coconut, condensed milk and vanilla. Chill. Shape into strawberries. Roll in sugar crystals. Mix food color with almonds to tint for stems. Makes about 48.

STRAWBERRY MOLDED CREAM

3-oz. pkg. unflavored gelatin
3/4 c. sugar
1 c. cream
1/2 pt. sour cream
2 c. sliced strawberries
1/2 c. sugar

Stir sugar and gelatin in double boiler. Add cream and stir over hot water until gelatin and sugar dissolve. Chill until mixture begins to set. Fold in sour cream. Turn into a mold and chill until set. Unmold; serve with strawberries mixed with sugar. Serves 6.

CHOCOLATE DIPPED STRAWBERRIES

6-oz. pkg. chocolate chips
2 T. butter
Whole strawberries

Melt chocolate chips and butter over hot water in double boiler. Stir. Dip whole strawberries in mixture until about 3/4 covered. Cool.

STRAWBERRY MOUSSE

1 egg white
1/8 t. salt
3/4 c. powdered sugar
1 c. whipping cream
1 1/2 c. mashed strawberries

Beat egg white and salt until foamy. Slowly add powdered sugar, beating until stiff. Whip cream. Fold berries into cream; fold into egg white mixture. Chill until firm.

STRAWBERRY-BANANA DESSERT

1 6-oz. pkg. strawberry gelatin
2 c. boiling water
2 10-oz. pkg. frozen strawberries
1 medium mashed banana
1 T. lemon juice
1 pt. sour cream

Dissolve gelatin in water; stir in berries. Mix in banana and lemon juice. Stir in sour cream. Pour into mold and chill until set.

STRAWBERRY PARFAIT RING

Strawberry sherbert
Vanilla ice cream

Soften sherbert and ice cream. Spoon sherbert on bottom of ring mold. Cover with layer of vanilla ice cream. Top with sherbert. Freeze.

STRAWBERRY SHERBERT

4 qts. sliced strawberries
4 c. sugar
2 2/3 c. milk
2/3 c. orange juice
1/8 t. ground ginger

Mix strawberries and sugar until juicy. Whir in blender. Blend in remaining ingredients. Pour into freezer trays. Freeze 2-3 hours, stirring 2-3 times. Pack in freezer containers. Makes about 1 gallon.

STRAWBERRY PUDDING

3 c. hulled strawberries
1 c. sugar
1/2 c. water
6 slices firm white bread
Butter

Cook strawberries, sugar and water. Butter bread. Alternate layers of bread and strawberry mixture in loaf pan. Chill. Slice and serve with whipped cream or ice cream.

STRAWBERRY ANGEL DELIGHT

3-oz. pkg. strawberry gelatin
1 c. boiling water
1 pt. vanilla ice cream
1 10-oz. pkg. frozen strawberries
1 small angel food cake

Dissolve gelatin in boiling water. Stir in ice cream until melted. Add strawberries and cake broken into small pieces. Chill until set. Slice in squares. Top with whipped cream.

STRAWBERRY-LEMON PARFAIT

Strawberry sherbert
Lemon sherbert

Spoon strawberry and lemon sherberts alternately into parfait glasses.

STRAWBERRY ICE CREAM

1 c. milk
1/3 c. sugar
2 T. flour
Dash salt
2 egg yolks
2 egg whites
1/3 c. sugar
1 qt. crushed strawberries
2 T. lemon juice
3/4 c. sugar
1 1/4 c. whipping cream

Stir milk, sugar, flour, salt and egg yolks over medium-high heat until mixture is custard-like. Chill well. Beat egg whites and sugar until stiff. Fold into custard. Mix together strawberries, sugar and lemon juice; set aside. Whip cream and add it to the strawberry mixture. Fold into custard. Pour into refrigerator trays. Freeze.

For a brochure describing other **eberly press** books, please write to:

eberly press
1004 Michigan Ave.
East Lansing, MI 48823